READING POWER

Man-Made Disasters

Poisoned Planet

Pollution in Our World

August Greeley

The Rosen Publishing Group's
PowerKids Press™
New York

Published in 2003 by The Rosen Publishing Group, Inc.
29 East 21st Street, New York, NY 10010

Copyright © 2003 by The Rosen Publishing Group, Inc.

All rights reserved. No part of this book may be reproduced in any form without permission in writing from the publisher, except by a reviewer.

First Edition

Book Design: Christopher Logan

Photo Credits: Cover © Larry Lee/Corbis; pp. 4-5, 14 © Sergio Dorantes/Corbis; pp. 4, 7, 8, 13, 15, 21 (globe) © PhotoDisc; p. 6 © Ted Spiegel/Corbis; p. 7 © Robert Landau/Corbis; pp. 8-9 © Charles O'Rear/Corbis; p. 9 (inset) © Alan Towse/Ecoscene/Corbis; pp. 10-11 © Stephanie Maze/Corbis; p. 11 (inset) © AP/Wide World Photo; p. 12 © Jenny Hager/The Image Works; p. 13 © David Reed/Corbis; p. 15 Christopher Logan; p. 16 © Charles & Josette Lenars/Corbis; p. 17 © Getty/FPG/Lester Lefkowitz; p. 17 (inset) © Joseph Sohm, Chromo Sohm Inc./Corbis; p. 18 © AFP/Corbis; p. 19 © Roger Ressmeyer/Corbis; p. 20 © Doug Wilson/Corbis; p. 21 © Index Stock Imagery Inc.

Library of Congress Cataloging-in-Publication Data

Greeley, August.
Poisoned planet : pollution in our world / August Greeley.
   v. cm. — (Man-made disasters)
Includes bibliographical references and index.
Contents: Pollution in our world — Air pollution — Acid rain — Water pollution — Stopping pollution.
ISBN 0-8239-6487-6 (library binding)
1. Pollution—Juvenile literature. [1. Pollution.] I. Title.
TD176 .G74 2003
363.73—dc21
                                           2002002935

Contents

Pollution in Our World	4
Air Pollution	8
Acid Rain	12
Water Pollution	14
Stopping Pollution	18
Glossary	22
Resources	23
Index/Word Count	24
Note	24

Pollution in Our World

Pollution makes the air, water, or land dirty. Almost anything can pollute our world. Pollution happens when solids, liquids, or gases get into the environment faster than the environment can turn them into something that is harmless.

Check It Out

Some chemicals that cause pollution can continue to harm Earth for 400 years.

Ever since groups of people started living together in one place thousands of years ago, pollution has been a problem. The more people there are in one place, the more waste there is.

Pollution is a big problem for many large cities.

Check It Out

Each person in the United States makes over four pounds of waste a day.

Air Pollution

Gases and toxic chemicals let out into the air from factories and car engines cause air pollution. Air pollution can make breathing difficult and can cause cancer and other diseases. Between 200,000 and 500,000 people die each year from the effects of air pollution.

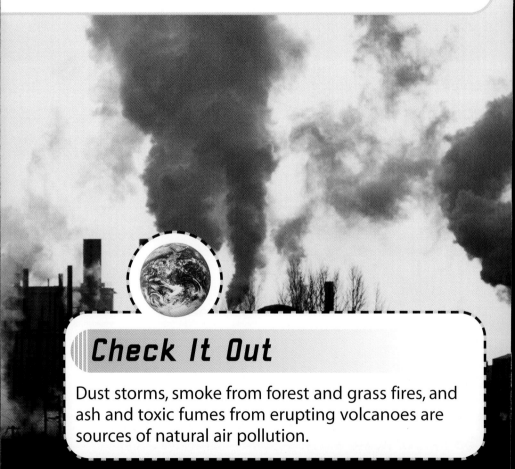

Check It Out

Dust storms, smoke from forest and grass fires, and ash and toxic fumes from erupting volcanoes are sources of natural air pollution.

Air pollution affects children more than adults. This is because children spend more time outdoors and their lungs are still growing. Many children get asthma from air pollution.

The fuels used to make gasoline for our cars and to heat our homes let out toxic fumes when they are burned. These fumes cause smog. Smog is a form of air pollution. Many cities, such as Los Angeles, Mexico City, and Tokyo, have problems with smog.

Each day, breathing the air in Mexico City is as bad for you as smoking forty cigarettes.

Police officers who direct traffic in Tokyo, Japan, are given clean oxygen at oxygen bars. People in Tokyo can also buy oxygen from machines at street corners.

Acid Rain

Sometimes, pollution in the air mixes with water vapor in the air. When this happens, acid rain occurs. Acid rain looks, feels, and tastes like clean rain, but it is harmful. When acid rain falls to Earth, it weakens the soil and causes trees and plants to grow slowly or die.

This forest in Pennsylvania was harmed by acid rain.

Acid rain that falls into lakes and streams hurts the water. Fish and other living things in or near the water are harmed, or even killed.

Check It Out

The pollution in acid rain can cause heart and lung problems in people.

Acid rain and other pollutants can destroy stone.

Water Pollution

In addition to acid rain, many other pollutants harm our water. Some factories dump toxic wastes underground. During storms, rainfall mixes with the waste.

Chemicals were dumped into this river in Mexico, turning the water white.

The polluted rainwater runs into the groundwater. The polluted groundwater can harm drinking water, lakes, rivers, and streams.

Check It Out

The drinking water for over half the people in the United States comes from groundwater.

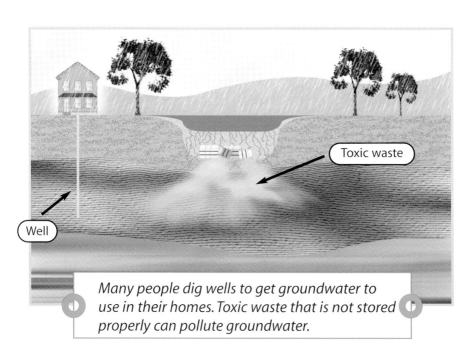

Many people dig wells to get groundwater to use in their homes. Toxic waste that is not stored properly can pollute groundwater.

Sometimes, trash and other waste is dumped directly into rivers, lakes, and oceans. This can harm many plants and animals. Animals can die if they swallow trash, such as plastic bags.

About 300,000 miles of rivers and streams in the United States are polluted.

Motor oil or household cleaners poured into the sewers can also get into our water and pollute it.

Trash and other pollutants that are poured into the sewers go into rivers, lakes, and streams.

Stopping Pollution

Governments around the world are working to stop pollution. In 2001, one hundred seventy-eight countries signed an agreement to lower the amounts of harmful gases they let out into the air. Some people are also trying to find new forms of energy that will not pollute the air, water, or land.

Many countries are worried about pollution in the environment.

- Some factories let out carbon. When carbon mixes with some gases in the air, the earth can be hurt. This can cause many problems, such as a rise in sea levels and flooding.

Countries that Let Out the Most Carbon into the Air

Country	Amount in Metric Tons
United States	1,495,000,000
China	740,000,000
Russia	405,000,000
Japan	288,000,000
Canada	139,000,000

Scientists are trying to find ways to make the energy that we get from the sun cheap enough for everyone to use.

You can help cut down on pollution, too. Do not litter. Turn off faucets and lights when you are not using them. Recycle cans, plastic, and glass bottles. There will be less waste and less pollution if we each do our part.

Recycling newspaper and other paper items is an easy way to help the earth.

Check It Out

One dripping faucet can waste 20 gallons of water a day!

Glossary

asthma (**az**-muh) an illness of the lungs that makes breathing difficult and can cause death

cancer (**kan**-suhr) an illness caused by a harmful growth of cells

diseases (duh-**zeez**-uhz) illnesses

engines (**ehn**-juhnz) motors that power machines

environment (ehn-**vy**-ruhn-muhnt) the air, water, and soil around us

faucet (**faw**-siht) the part of a pipe that controls the flow of water

oxygen (**ahk**-suh-juhn) a colorless, odorless gas that we breathe in

pesticides (**pehs**-tuh-sydz) chemicals that are used to kill pests, such as insects

pollutants (puh-**loot**-nts) wastes, chemicals, and gases that are put into the air, water, or soil

pollution (puh-**loo**-shuhn) anything that makes the environment dirty

recycle (ree-**sy**-kuhl) to make old items ready for use again

smog (**smahg**) a smoky fog caused by pollution

Resources

Books

Air and Energy: How We Use and Abuse Our Planet
by Arthur Haswell and Pamela Grant
Thameside Press (2000)

Pollution: Problems & Solutions
by National Wildlife Federation
McGraw-Hill (1998)

Web Sites

Due to the changing nature of Internet links, PowerKids Press has developed an online list of Web sites related to the subjects of this book. This site is updated regularly. Please use this link to access the list:

http://www.powerkidslinks.com/mmd/pppw/

Index

A
acid rain, 12–14

C
cancer, 8

D
diseases, 8

E
engines, 8
environment, 4, 18

F
faucet, 20–21

P
pollutants, 13–14, 17
pollution, 4, 6, 8–10, 12–13, 18, 20

S
smog, 10

W
waste, 6–7, 14–16, 20

Word Count: 509

Note to Librarians, Teachers, and Parents

If reading is a challenge, Reading Power is a solution! Reading Power is perfect for readers who want high-interest subject matter at an accessible reading level. These fact-filled, photo-illustrated books are designed for readers who want straightforward vocabulary, engaging topics, and a manageable reading experience. With clear picture/text correspondence, leveled Reading Power books put the reader in charge. Now readers have the power to get the information they want and the skills they need in a user-friendly format.